Work 137

黑山羊，不是替罪羊

Black Goat, not Scapegoat

Gunter Pauli

冈特·鲍利 著

凯瑟琳娜·巴赫 绘
高 青 译

丛书编委会

主　任：贾　峰

副主任：何家振　闫世东　郑立明

委　员：牛玲娟　李原原　李曙东　李鹏辉　吴建民
　　　　彭　勇　冯　缨　靳增江

特别感谢以下热心人士对译稿润色工作的支持：

王必斗　王明远　王云斋　徐小怗　梅益凤　田荣义
乔　旭　张跃跃　王　征　厉　云　戴　虹　王　逊
李　璐　张兆旭　叶大伟　于　辉　李　雪　刘彦鑫
刘晋邑　乌　佳　潘　旭　白永喆　朱　廷　刘庭秀
朱　溪　魏辅文　唐亚飞　张海鹏　刘　在　张敬尧
邱俊松　程　超　孙鑫晶　朱　青　赵　锋　胡　玮
丁　蓓　张朝鑫　史　苗　陈来秀　冯　朴　何　明
郭昌奉　王　强　杨永玉　余　刚　姚志彬　兰　兵
廖　莹　张先斌

目录

黑山羊，不是替罪羊	4
你知道吗？	22
想一想	26
自己动手！	27
学科知识	28
情感智慧	29
艺术	29
思维拓展	30
动手能力	30
故事灵感来自	31

Contents

Black Goat, not Scapegoat	4
Did you know?	22
Think about it	26
Do it yourself!	27
Academic Knowledge	28
Emotional Intelligence	29
The Arts	29
Systems: Making the Connections	30
Capacity to Implement	30
This fable is inspired by	31

牧羊人骑着马跟在一群山羊后面。一只长耳朵小跳鼠正在看着这一大群黑山羊走过。

"好像你们现在的数量比几年前多了。"它对一只山羊说。

"哦，是的，牧羊人想要更多我们的毛发，所以他的羊群翻了一番。"山羊回答说。

A herder is following his flock of grazing goats on horseback. A jerboa, a small mouse with very long ears is watching as a family of big, black goats walk by.

"It seems that there are many more of you now than a few years ago," he remarks to one of the goats.

"Oh yes, our herder wants more of our hair so he has doubled the size of our flock," Goat replies.

牧羊人跟在一群山羊后面……

A herder is following his flock...

最抢手的是我们的羊毛……

Most sought after for our wool...

"你确定他只对你的毛发，而不是对你的肉或奶感兴趣吗？"跳鼠问。

"你不知道因为我们的羊毛，我们黑山羊成了最受追捧的明星了吗？人们用它做出温暖的毛衣、披肩和外套。"

"对不起，不过我现在很困惑，你是一只山羊，对吗？我以为只有用绵羊身上的羊毛，才能给人们做出温暖的衣服。"

"Are you sure he is interested only in your hair, not your meat or milk?" Jerboa asks.
"Don't you know that us black goats are the most sought after, for our wool? People make warm sweaters, shawls and coats from it."
"Excuse me, but now I am confused. You are a goat, right? I thought it was sheep that gave wool, to help people dress warmly."

"是的,但他们的毛发被称为羊毛,而我的毛发被称为羊绒。"

"羊……绒?那听起来好像更值钱!"

"哦,那当然!人们愿意付出更多的钱,购买我们柔软细腻的毛发,牧民们剪毛时也会非常小心,在织成细纱前会轻轻地把它们洗净。"山羊自豪地说道。

"That is true, but their hair is called wool and my hair is called cashmere."

"Cash-mere? That sounds like there is more money in it!"

"Oh yes, there is. People are willing to pay so much more for our soft, fine hair that the herder shears it with great care and then washes it gently before spinning it into a fine yarn," Goat says proudly.

羊绒!

Cashmere!

几乎不够涵盖他的成本……

Hardly covers his costs ...

"现在我明白为什么牧羊人想拥有更多的山羊了。蒙古大草原放牧是免费的,牧羊人可以存下所有用你的毛发挣来的钱。"

"你先别下结论,跳鼠先生。农民可以自由进入土地,这是事实,但是中间商支付给他的辛苦钱,几乎不够涵盖他的成本。"

"这太不合理了!"跳鼠大声说,"没有你的毛发,和他的剪切、洗涤和纺织,哪来的毛衣呢!"

"Now I understand why the herder wants to keep more of you. The grazing is free all across Mongolia, so he gets to keep all the money he makes from your hair."

"Now don't you jump to conclusions, Mr Jerboa. It is true that the farmer has free access to land, but what he is paid by the middleman for all his hard work, hardly covers his costs."

"That just does not make any sense!" Jerboa exclaims. "Without your hair, and his shearing, washing and spinning, there would be no sweater."

"你说得非常对,但这就是现代商业——那些从生产者那里买进再卖出的商人才是最赚钱的人,生产者几乎很难生存,还必须不惜一切代价降低成本,生产更多的产品。"

"这怎么可能?"跳鼠想知道原因。

"You are so right. But this is modern day business for you – the one who buys from the producer and sells it on makes the most money. The one who produces barely survives, and has to cut costs at all cost, producing more and more."

"How is this possible?" Jerboa wants to know.

买进再卖出的商人才是最赚钱的人

The one who buys and sells makes the money

为了生存需要更多的山羊……

Needs more goats just to survive...

"这还不是最终的结果！情况变得越来越糟糕了。连从事电子支付的人都和我的牧羊人挣得一样多，可牧羊人的一生都致力于照顾我们。"山羊叹了口气。

"所以你的牧羊人需要更多的山羊？是为了生存吗？"

"是的，在当今的商业模式中，人们赚钱是通过买卖物品赚差价，而不是为了生产。"

"And that is not the end of it! It gets worse. The person who arranges payment over computer makes as much money as my herder does! And his entire life is dedicated to caring for us," Goat sighs.

"So that is why your herder needs to have more goats? Just to survive?"

"Yep. In business today, people make money by getting commission on selling items, not for producing them."

"听上去这似乎不公平!还有一些让我烦恼的事情。"跳鼠补充说,"如果每个牧羊人增加他在公共土地上吃草的山羊数量,那么很快给它们的食物就不够了。"

"这就是'看不见的手':每个人都有自己的利益,结果,我们会有太多的人到处去放牧,整个地区很快就会超负荷放牧,草原变为逐渐扩张的沙漠,这该怪谁呢?"

"That does not seem fair! And, there is something else that bothers me," Jerboa adds. "If every herder increases the number of goats he grazes on public land, soon there will not be enough food for them all."

"It is 'the invisible hand': everyone has his or her own profit in mind. As a result, they will all have too many of us grazing everywhere. Soon the entire area becomes overgrazed, and so the desert expands. And who will they blame?"

整个地区很快就会超负荷放牧

The area will become overgrazed

我因为创造出的沙漠而受指责

I will be to blame for creating the desert

"但是，如果牧羊人无法生存而最终所有的土地都变成了沙漠的话，想象一下羊绒到时候会多么昂贵呀……"

"牧羊人会放弃生意，像许多人一样到城里生活。我会成为替罪羊，因为创造出的沙漠而受指责！然后这里仅存的山羊群将成为博物馆里的……"

"But if the herder cannot survive, and in the end all the land is turned into desert, imagine how expensive cashmere will then be…"

"The herder will be out of business and go live in the city, like so many others. And I will be the scapegoat, the one they will blame for having created the desert! The only goats around here will then be those in the museum…"

"这同样不公平!每个人都应该赞美你,为你的美好、温暖、自然的毛发,也应赞美你的牧羊人为你做出的一切辛勤工作。他和你都理应过上幸福的生活。"

……这仅仅是开始!……

"That is simply not fair either! Everyone should celebrate you for your fine, warm, natural hair, and your herder for all his hard work. He deserves making a good living from farming goats so he, and you, can lead happy lives."

... AND IT HAS ONLY JUST BEGUN!...

……这仅仅是开始！……

...AND IT HAS ONLY JUST BEGUN!...

Did You Know?
你知道吗？

The economist Adam Smith first described the 'invisible hand', in 1776. He stated that if all go after their own profit, then there will be social benefits for society.

经济学家亚当·斯密在1776年首次描述了"看不见的手"，他指出如果所有人都追求自己的利润，那么社会就会有社会效益。

Economist William Foster Lloyd was the first to debate the 'tragedy of the commons', in 1833. He stated that when individuals pursue only their own interest using a commons, they will end up depleting that resource altogether.

经济学家威廉·福斯特·劳埃德在1833年最先提出"公共地悲剧"。他表示，当个人只顾用公共资源追求自己的利益时，最终他们就会完全消耗这些资源。

Commons are resources available to all members of society and includes clean air, fresh water, fertile soil, forests and city parks. These assets are held in common, and are not owned privately.

公共资源是社会所有成员的资源，包括新鲜空气、淡水、肥沃的土壤、森林和城市公园。这些资产是共有的，不属于私有财产。

Nature offers many services for free: soil to produce food and rain to provide water, regulating climate and controlling disease, cycling nutrients through water, air and soil, and a natural environment that inspires us and offers recreation.

自然提供许多免费服务：土壤产生食物；雨水提供水；调节气候和控制疾病，通过水、空气和土壤循环营养；自然环境供我们休养生息。

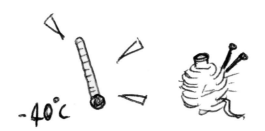

Goats only produce cashmere when the winters are cold (up to -40° C) and the summers in the Gobi desert are extremely hot (up to +40° C). The goat herder accompanies his animals everywhere and at all times.

当冬季气温降至零下40摄氏度，或当夏季戈壁沙漠中气温升至40摄氏度的高温时，山羊才生产羊绒。牧羊人们却是随处随地陪伴着他的动物。

When the economic model prescribes the production of more, then raw material supply must increase to keep up with demand. Since natural resources are limited, going beyond the limit means destroying supply.

经济模式规定的生产量增加时，原料供应量必须增加以跟上其需求。由于自然资源有限，超出限制意味着破坏供应。

The carrying capacity is the number of goats or cows that an ecosystem can sustain indefinitely, when availability of food and water, and natural regeneration, are taken into account.

承载能力是指当食物和水的可用性、自然的再生能力被考虑在内时，一个生态系统可以无限期地支撑的山羊或奶牛的数量。

Cashmere is like an undercoat or 'fleece jacket' of very soft hair. Goats also have an overcoat with coarse hair. An estimated 150 million goats live in and around the Gobi desert.

羊绒就像一件非常柔软的"羊毛夹克"。山羊也有一件毛毛的大衣。据统计，约有1.5亿只山羊住在戈壁沙漠及其周围。

Think About It

想一想

What is the carrying capacity of the house in which you live? How many people can live comfortably with the space, water and power you have available?

你住的房子的承载能力是多少？有多少人能够舒适地享受你所拥有的空间、水和电？

Do you think there is 'an invisible hand' guiding the economy so that everyone will benefit?

你认为有一只"看不见的手"在指导经济，让每一个人都能受益吗？

If everyone uses the grassland to graze their goats, what would happen if everyone increased their number of goats by adding one extra goat? And then by adding two? And then three…?

如果每个人都在牧场放牧山羊，那么如果每个人都增加一只山羊将会发生怎样的变化呢？再加两个？然后三个？

Do you agree that the company that arranges electronic payment (over the internet) could earn as much as the goat herder or farmer? Should this be regulated?

你是否同意电子支付公司（通过互联网）可以赚取与牧羊人或农民一样多的收入？这应该受到管制吗？

Have you ever touched a cashmere garment? Find a store that sells cashmere items, like sweaters or shawls. Feel the texture of the cashmere and how different it feels to wool. Now have a look at the price. Does it seem expensive? How much cheaper would the same garment be in wool or cotton? Find five or six similar garments made from other fabrics and create a brief inventory of garment and price. Now ask around and find out why there is such a big difference in price. Calculate how much each person in the cashmere production chain earns from it: the herder, the one who shears, the one who washes it, the one who spins it, the one who weaves it … up to the one who sells it in the store. How does the price paid in the shop compare with how much the herder, who spends all summer and winter with the goats, got for it? Ask yourself: would you like to be a goat herder at that price?

你有没有摸过羊绒衣服？找一家销售羊绒制品如毛衣或披肩的商店。感受羊绒质感以及和羊毛触感的差别。再看看价格。贵不贵？同样的服装羊毛的或棉质的会便宜多少？找其他面料制成的五六件类似的服装，并创建一个简单的服装和价格清单。询问周围的人，找出价格差异这么大的原因。计算羊绒生产链中每个人从中赚取的收益：牧羊人，剪羊毛的人，洗羊毛的人，纺羊毛的人，编织的人……直到在商店销售的人。在店里支付的价钱如何与整个夏天和冬天都陪伴着山羊的牧羊人所赚的钱相比较呢？问问自己：这样的价格，你愿意成为一名牧羊人吗？

TEACHER AND PARENT GUIDE

学科知识
Academic Knowledge

生物学	承载能力；生态系统服务，如水、土壤、卫生、循环；山羊春季蜕毛；长耳跳鼠。
化 学	羊毛主要由蛋白质组成；羊毛上的一种分泌油脂被称为羊毛脂。
物 理	羊绒吸收水的能力消除了湿气；穿羊毛制品的人通过促进蒸发传热而调节体温；使用羊毛作为绝缘材料；羊绒的标准直径不超过19微米。
工程学	纺纱锭子的发展。
经济学	"看不见的手"；"公共悲剧"，免费资源如何被过度利用；运用中介机构的成本以及通过互联网提供支付安全的成本。
伦理学	通过一个更便宜的方式产生更多的产品的经济，会导致成本的外化，而这些成本将由整个社会去承担，如污染空气、退化的土地、受污染的水等等。每个人如何在产品的价值链中上获取利益，产品的销售价格取决于供给和需求的相互作用。
历 史	绵羊和山羊在一万年前就被驯化了；罗马人在英国发现了羊毛行业，并刺激其进一步发展。
地 理	戈壁沙漠；羊绒曾在克什米尔地区加工并大量销往欧洲，因此得名。蒙古为所有山羊牧民提供免费进入牧场的土地；长耳跳鼠的栖息地。
数 学	在价格之上的佣金，在基本固定成本之上增加百分比；基本百分比方程；涨价和利润率。
生活方式	我们寻求低价格的优质产品，往往不知道的供应链是通过对价值链底部生产商的成本降低；飞船上的宇航员穿羊毛以求舒适。
社会学	社会公益的重要性和作用；基本活动如捕鱼，放牧，农业和林业的作用。
心理学	预期的重要性；形成负面或积极的心态；指责别人团体的内部和外部的替罪羊效应。
系统论	所有人都可以使用公共资源，但过度使用会导致由公共资源提供的生态服务基础的破坏。

教师与家长指南

情感智慧
Emotional Intelligence

长耳跳鼠

跳鼠是善于观察的,它注意到山羊数量的变化从而怀疑牧羊人的动机。起初它不了解山羊的逻辑,也不了解"羊绒"一词的含义时,它说出它的不确定并去询问确认。跳鼠很快掌握了牧羊人所获得的优势,并认为牧羊人只是为自己赚更多的钱。随着它收到的信息更多,跳鼠变得更加困惑。它有信心去澄清更多的问题。它应用逻辑得出的结论是,目前的商业模式没有解决方案,但希望有一种方式可以让大家开心。

黑山羊

黑山羊直接而准确地回答了跳鼠的问题。它有自我意识,知道它毛发的高价值。它表现出自豪感,并赞赏牧羊人对自己的毛发进行纺纱的方式。它为牧羊人的行为辩护,并要求跳鼠不要如此迅速地得出否定结论。它仔细研究了周围的情况,认为牧羊人不应该对此负责。当跳鼠的问题揭示了一个残酷的现实时,黑山羊显示出它的透彻。山羊说预测是相当负面的。它知道,简单地忽略事实可能导致牧场荒漠化。它还预见到在未来的某一个时候,仅存的"羊绒"山羊群会被发现在博物馆里。

艺术
The Arts

让我们看看我们可以用羊毛做什么,这是非常强大的天然材料。其独特之处在于它的重复使用性。一旦毛衣失去了原有的形状,或手套上有了一个洞,就可以重新编织它,使其成为新的东西。再次使用羊毛的艺术在于方式可以各种各样。唯一的限制是你的想象力!你想用它钩编,还是把它拼接到被子上或其他织物艺术品上?你可以使用许多剩余的纱线,或者将爷爷的旧毛衣分开来编制出新的东西,并将你独特的标志放在上面。

TEACHER AND PARENT GUIDE

思维拓展
Systems: Making the Connections

近来对羊绒的需求不断增长，但其供应有限是由自然因素决定的。穿羊绒衫会带来美好的形象，所以它是一个非常受欢迎的产品。因此，超市希望以低成本提供该产品。如今新型的超轻和薄纺纤维都以一个非常低的价格在出售。价格低廉至似乎无法让任何人从中获利。这种低价格和可用性广泛的产品提高了人们的期望值和更多的需求。问题在于，随着市场对羊绒需求的不断增长，唯一的办法就是在有限的土地上饲养越来越多的山羊。动物数量以三倍的速度增长将直接导致牧场的灭绝。这意味着山羊养殖和羊绒生产的彻底破坏。不幸的是，越来越多的山羊和更高的羊绒产量并没有转化为额外的收入和利润。这意味着即使山羊数量增多，牧民的收入也不会增加，所以问题是："谁从生态系统的破坏中赚钱？"羊绒的生产者得到的价格很低，甚至支付不起他子女的教育费用。超市和奢侈品商店对这些奢侈品进行标价。最终价格中，牧民所占的比例只有百分之几。这是对牧民生存的极大挑战。唯一的解决方案是让牧民赚更多的钱，有了更多的钱，他们就会减少山羊的数量。建议消费者同意向牧民支付最终价格的10%。这将产生巨大的影响。由于从消费者那里获得直接佣金，牧民得到更高的价格，那么他的收入就会增加。一旦收入大幅增加，他将能够减少山羊的数量，而且仍然比以前赚得更多。

动手能力
Capacity to Implement

你是一个好的销售人员吗？羊绒总是需求很高。每个人都想拥有一件价格优惠的羊绒服装。问题是生态系统脆弱，产量增加将导致生态系统遭到破坏。建立一个支持特殊税收的论证，即消费者自愿支付此税收。这个税收就像增值税或销售税，每个人都同意这些税收将支付给牧民。当你解释了这些影响，测试一下这些潜在客户的反应。验证这样一个大胆的提案是否有意义。坚持要立即实现这一点，因为戈壁沙漠正在迅速扩大，山羊将会被留在没有牧场的地方。山羊可能会成为博物馆的展品！

教师与家长指南

故事灵感来自

绍伦高·巴特别克赫
Solongo Batbekh

绍伦高·巴特别克赫是位于乌兰巴托（蒙古）的 Goyo 羊绒公司执行董事。她曾在蒙古科技大学学习科学与工程。在获得银行业务和饮料业务工作的经验后，她决定将自己的事业发展到围绕戈壁沙漠的可持续的羊绒行业里。她创建了牧民的第一个合作社，目的是生产有机羊绒，为牧民创造更好的收入。她寻找一个有助于保护生态系统的企业，同时确保牧民获得体面的生活。

图书在版编目（CIP）数据

黑山羊，不是替罪羊：汉英对照／（比）冈特·鲍利著；（哥伦）凯瑟琳娜·巴赫绘；高青译．— 上海：学林出版社，2017.10
（冈特生态童书．第四辑）
ISBN 978-7-5486-1261-2

Ⅰ．①黑… Ⅱ．①冈… ②凯… ③高… Ⅲ．①生态环境－环境保护－儿童读物－汉、英 Ⅳ．① X171.1-49

中国版本图书馆 CIP 数据核字（2017）第 143522 号

© 2017 Gunter Pauli
著作权合同登记号　图字 09-2017-532 号

冈特生态童书
黑山羊，不是替罪羊

作　　者——	冈特·鲍利
译　　者——	高　青
策　　划——	匡志强　张　蓉
责任编辑——	胡雅君
装帧设计——	魏　来
出　　版——	上海世纪出版股份有限公司 学林出版社
	地　址：上海钦州南路81号　电话／传真：021-64515005
	网　址：www.xuelinpress.com
发　　行——	上海世纪出版股份有限公司发行中心
	（上海福建中路193号　网址：www.ewen.co）
印　　刷——	上海丽佳制版印刷有限公司
开　　本——	710×1020　1/16
印　　张——	2
字　　数——	5万
版　　次——	2017年10第1版
	2017年10月第1次印刷
书　　号——	ISBN 978-7-5486-1261-2/G.487
定　　价——	10.00元

（如发生印刷、装订质量问题，读者可向工厂调换）